AF228479

Organic Gems

by Grace Hansen

GEOLOGY ROCKS!

Abdo Kids Jumbo is an Imprint of Abdo Kids
abdobooks.com

abdobooks.com

Published by Abdo Kids, a division of ABDO, P.O. Box 398166, Minneapolis, Minnesota 55439.
Copyright © 2020 by Abdo Consulting Group, Inc. International copyrights reserved in all countries.
No part of this book may be reproduced in any form without written permission from the publisher.
Abdo Kids Jumbo™ is a trademark and logo of Abdo Kids.

Printed in China

052019

092019

 THIS BOOK CONTAINS RECYCLED MATERIALS

Photo Credits: Alamy, Depositphotos Enterprise, iStock, Shutterstock

Production Contributors: Teddy Borth, Jennie Forsberg, Grace Hansen
Design Contributors: Dorothy Toth, Pakou Moua

Library of Congress Control Number: 2018963348
Publisher's Cataloging-in-Publication Data

Names: Hansen, Grace, author.
Title: Organic gems / by Grace Hansen.
Description: Minneapolis, Minnesota : Abdo Kids, 2020 | Series: Geology rocks!
    set 2 | Includes online resources and index.
Identifiers: ISBN 9781532185601 (lib. bdg.) | ISBN 9781532186585 (ebook) |
    ISBN 9781532187070 (Read-to-me ebook)
Subjects: LCSH: Gems--Juvenile literature. | Precious stones--Juvenile literature. |
    Rocks--Identification--Juvenile literature. | Geology--Juvenile literature.
Classification: DDC 553.8--dc23

Table of Contents

Organic Gems

All gems have three things in common. They are all strong, rare, and beautiful.

Organic Gems are different from crystal gems. Organic gems can also be called living gems. This is because they form from living or once living things.

Organic gems can form on land and at sea.

Organic Gems at Sea

Pearls form at sea. A pearl
forms inside a **mollusk**,
such as an oyster.

When a small material enters an oyster, the oyster forms shiny layers around it. This shiny material is called **nacre**. The layers build up until a beautiful pearl is formed.

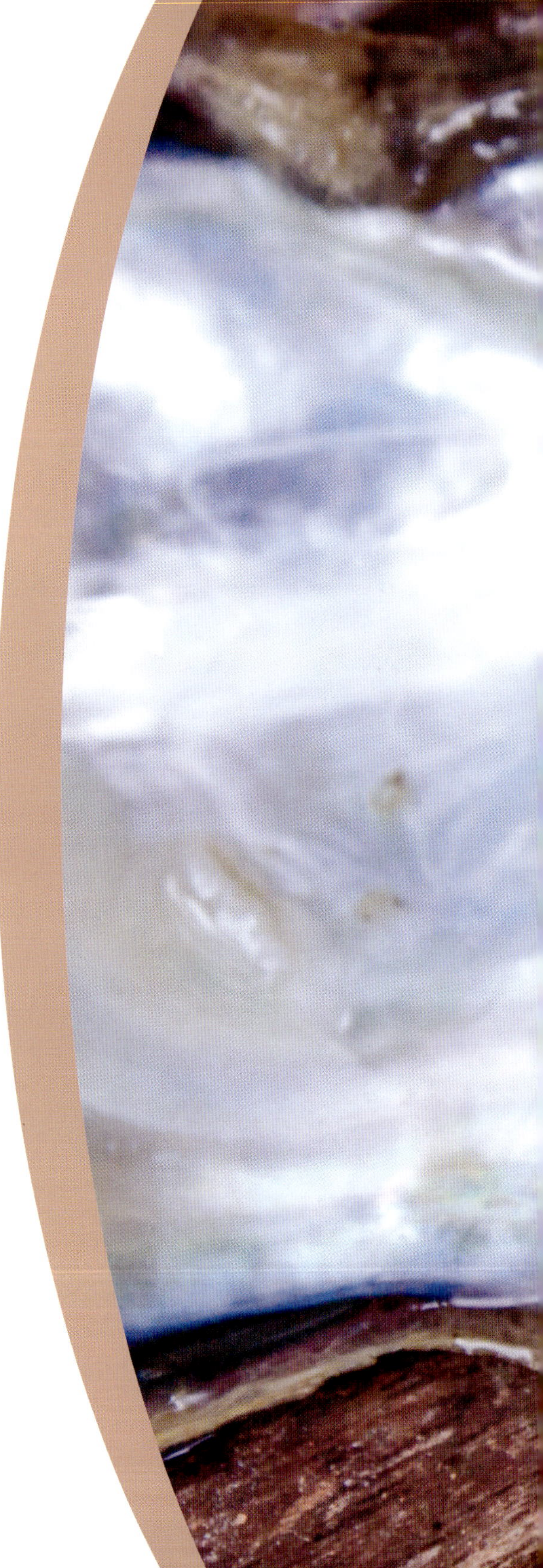

13

Coral also forms at sea. Coral
may look like a plant, but it is
a living animal. This animal has
a skeleton. When it dies, its
skeleton is left behind.

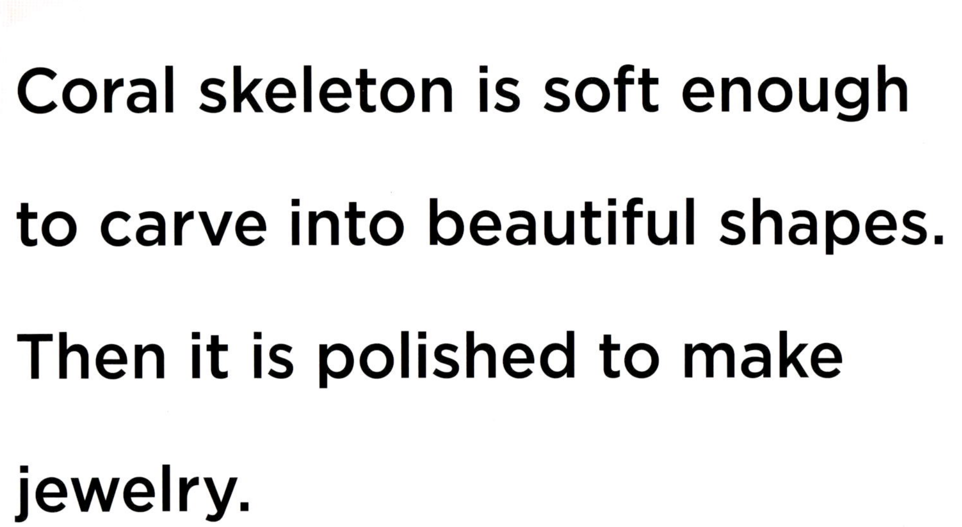

Coral skeleton is soft enough
to carve into beautiful shapes.
Then it is polished to make
jewelry.

Organic Gems on Land

Some organic gems were made from **ancient** plants. So, they are also **fossils**! Amber was made from the resin of ancient pine trees.

Jet is wood that decayed under extreme pressure. This wood is from trees that stood millions of years ago. Dinosaurs likely ate leaves from these trees!

Common Organic Gems

Glossary

ancient – very old.

fossil – the remains or trace of a living animal or plant from a long time ago.

mollusk – an animal in a large group of invertebrates. Most kinds live in the ocean and have soft bodies covered by a shell. Clams, oysters, and snails are examples.

nacre – a hard, iridescent material lining certain seashells, such as that of the pearl oyster. It is also called mother of pearl.

Index

Abdo Kids ONLINE

FREE! ONLINE MULTIMEDIA RESOURCES

Visit **abdokids.com** to access crafts, games, videos, and more!

Use Abdo Kids code **GOK5601** or scan this QR code!